# REPORT,

## HISTORICAL AND STATISTICAL,

ON THE COLLECTIONS IN

# Geology, Zoölogy and Botany

IN THE MUSEUM OF THE

# UNIVERSITY OF MICHIGAN,

**Made to the Board of Regents, Oct. 2d, 1863,**

BY ALEXANDER WINCHELL, A. M.,

PROF. GEOL., ZOOL. AND BOTANY.

---

ANN ARBOR:
PUBLISHED BY THE UNIVERSITY,
1864.

ELIHU B. POND,
*PRINTER,*
OFFICE OF THE MICHIGAN ARGUS.

# REPORT.

To the Honorable, the Board of Regents:

The undersigned regards the present a suitable occasion for bringing together, in a summary manner, such facts as are accessible relative to the history and growth of those portions of the museum which are placed under his particular supervision, and for furnishing an approximate estimate of the extent of our means of illustration in the various departments of Natural History which he has the honor to represent.

The early history of the museum is preserved only in tradition. There are no records dating back anterior to 1855, save the official records of appropriations made by the Boards of Trustees and Regents. The first professor elected in the Literary Department was an incumbent of the chair of Zoology and Botany. This was Asa Gray, M. D., elected in 1838, but never called to active service; and for many years past the distinguished "Fisher Professor of Natural History" in Harvard University. During the previous year, 1837, Douglass Houghton, M. D., and Abram Sager, M. D., had been conducting the newly organized survey of the Natural History of the state. This had been instituted when the state was yet but one year old. The original constitution of the survey provided that the specimens should be deposited in the University.* The nucleus of the museum of Natural History, however, was the Lederer Collection of minerals purchased in 1838. To this began soon to be added the collections of the scientific survey. Dr. Houghton was elected, in 1839, Professor of Chemistry and Mineralogy; and in 1842, Dr. Sager, already a professor in the Medical Department, was elected to

*Laws of Michigan 1837, p. 14.

the chair of Botany and Zoology, which he held till 1850, and the duties of which he continued to perform till the appointment of the present incumbent, in 1855. After the death of Dr. Houghton, and the suspension of the survey, in 1845, the legislature, by a special enactment, secured the University in the possession of all the collections of the survey.*

The materials of the survey prosecuted under the direction of Dr. Houghton, consisted as follows:

1. *Geological Specimens.* The number of these was large, and they represented all portions of the state. They embraced however but a limited number of fossils and most of these were in an imperfect state of preservation. The lithology of the state was well illustrated; but, as in all surveys, multitudes of rock specimens were collected, merely to answer a temporary use in a final investigation—serving rather as the memoranda of tacts to be finally incorporated, than as the materials of a showy, or even a very valuable cabinet of specimens The paucity of fossils in this collection is naturally attributable to two good causes: *first*, the remarkable fewness of fossiliferous outcrops, especially at that period in our municipal history, and *second*, the nature of the methods by which surveys were prosecuted at that stage of scientific development. The ores of the collection are attached to the Mineralogical Cabinet. About 2000 of these geological specimens are at present lying on the shelves of the museum, and about 1000 duplicates are packed away.

2. *Zoological Specimens.* These were collected under the direction of Dr. (now Professor) Abram Sager, and have received from him a large amount of accurate investigation. They embrace about 25 entries of mammals, 241 of birds, represented by about 750 specimens, 50 of reptiles and batrachians, 50 of fishes, besides a small number of Articulates. The collection of birds embraced nearly all the species which visit our state.

3. *Botanical Specimens.* These were collected by Dr. Sager in 1838, Dr. Wright and his assistant in 1839, and by Dr. Houghton and an assistant during the subsequent years. This herbarium is found to contain specimens of 900 species of Michigan plants, many of which exist in numerous duplicates.

---

*Laws of Michigan, 1846, p. 199.

The principal accessions to the museum from later sources, are the following, arranged as nearly as practicable in the order in which the persons named began their contributions.

DR. WESLEY NEWCOMB, of Albany, N. Y., presented the University, in May, 1843, at the instance of Dr. Houghton, 82 species of shells, of which 54 species were American land shells, one a European land shell, and the remainder marine species.

PROF. DOUGLASS HOUGHTON, M. D. After the death of Dr. Houghton his scientific Library and private cabinet of Natural History were purchased by the University. This purchase (besides minerals) furnished a limited number of geological specimens and a considerable number of marine shells.

ASHER B. BATES, ESQ., of Honolulu, S. I. presented some shells from the south Pacific, together with a few samples of lava from Hawai, and an autograph letter of King Liho-liho. Mr. Bates also donated sundry articles of south sea manufacture which possess considerable ethnological interest.

PROF. ZINA PITCHER, M. D., presented incomplete skeletons of a manatus, a grampus, an alligator, and an African monkey. A specimen of flying fish (*Dactylopterus*) in the museum is also supposed to have been presented by him.

PROF. ALVAH BRADISH presented the skin of an alligator from Georgia and skins of sundry fishes from the West Indies.

T. R. CHASE, ESQ., of Cleveland, Ohio, an alumnus of the University, presented a fine collection of coal-plants from the coal mines of northern Ohio; to which he added in June 1863, a small lot of fossils, finely preserved, from Kelly's Island, Lake Erie.

PROF. I. A. LAPHAM, of Milwaukee, furnished the museum with a good labeled assortment of *Unionidae* from the Ohio and Mississippi rivers.

JOSEPH MONDS, ESQ., sold the Board of Regents in 1855 a collection of shells embracing nearly 2000 entries and about 6000 specimens. The specimens represent the New England coast, the West Indies, the East Indies, the Chinese seas, Madagascar, and other shores. The air breathing mollusca are beautifully and extensively represented in this collection by the principal species of Europe, South America and the East Indies. The beautiful family of Cowries is also very abundantly illustrated. The purchase embraced, also, several excellent specimens of corals, echinoderms and crustacea. The collection had been accumulated during a residence of many years at Salem, Mass. from the whalers returning to that port. Its cost to the University was $400. Mr. Monds subsequently donated two or three dozen specimens of shells.

This purchase constitutes the principal bulk of our conchological cabinet. It is a misfortune that not a few of the specimens lack their specific names; while the library is not in possession of such works as are indispensable for their identification.

PROF. ABRAM SAGER, M. D., has given the museum from time to time, sundry valuable articles consisting of reptiles, birds, insects and geological specimens—among the latter a magnificent specimen of *Syringopora* from this State.

PROF. A. WINCHELL has continued to contribute to the museum from 1855 to the present. The donations embrace specimens in the principal classes of Vertebrates, Articulates and Molluscs as well as specimens in Geology. When certain purposes—still somewhat contingent—shall have been carried into execution, the donations from this source will become worthy of a special communication.

J. DELASKI, M. D., presented the museum in 1857 a young *Boa Constrictor* from Brazil, (besides sundry rare geological reports), and in 1860, sent from the coast of Maine, a collection of marine fishes, in alcohol, together with a quantity of shells, some geological specimens and other objects of natural history.

D. M. JOHNSON, a student, of Coshocton, Ohio, presented in 1857, and at subsequent times, a number of Ohio shells, some specimens from the coal measures of that State, and objects of interest from other localities.

W. J. BEAL, of Rollin, Lenawee County, has donated large amounts of zoological specimens, representing most of the classes. In 1857–8–9 the museum received from him sundry specimens of reptiles of infrequent occurrence in this State, besides some skins of birds. In May and June 1863 Mr. Beal, while a student under Agassiz at Cambridge, sent us two boxes containing about 1700 zoological specimens and 163 entries authentically labeled, from the sea coast of New England.

George E. Cumming of LaFayette, Ind., donated, in 1857, several reptiles from that State.

Aaron C. Jewett, late of Ann Arbor, and an honored and lamented graduate of the University, has donated from time to time, reptiles, fresh water shells, corals and Indian relics.

E. W. McGraw collected in Ann Arbor, in 1857–8, a large number of *Salamandridæ* which are deposited in the museum.

T. A. McGraw increased the collections of *Salamandridæ* by further additions from Ann Arbor.

S. H. White an alumnus, and now of Chicago, presented in 1857, 64 species of fossils, authentically labeled, mostly from the Clinton and Niagara groups in the vicinity of Lockport, New York.

Smithsonian Institution. The first donation from this Institution was one in 1858, of 75 species of reptiles in alcohol, authentically labeled. These were subsequently incorporated in the "Trowbridge Collection" by request of the Smithsonian Institution.

In 1859 the museum received the specimens of the Trowbridge Collection.

In 1861 it received 127 entries of the skins of birds from Arctic America. These were in part from the collections made by Mr. Robert Kennicott under the auspices of the Smithsonian Institution, the Audubon Club, and the University of Michigan—the latter having contributed $200 toward Mr. Kennicott's expenses. It may be proper, even here to add that the objects of Mr. Kennicott's expedition were most materially promoted by the facilities freely afforded by the Hudson's Bay Fur Company.

In July 1861, the museum received from the Smithsonian Institution 204 entries of marine shells from the Indo-Pacific.

In 1863 a further installment of skins from the Arctic zone was received, together with an extremely rare collection of the eggs of Arctic birds.

In June, 1863, the Institution forwarded a collection of *Unionidæ*, consisting of 81 entries. These were mostly labeled by numbers referring to the Smithsonian "Check Lists," but unfortunately most of the localities were wanting.

A. H Castle, an alumnus, presented, in 1858, a good collection of fossils from the "Waukesha limestone" of Wisconsin.

Lieut. W. P. Trowbridge, late a professor in the University, deposited in 1859, the "Trowbridge Collection" of Zoology, which, in 1861, was unconditionally donated to the University. Lieut. Trowbridge, having, while employed on the United States Coast Survey in California, devoted a large amount of attention to the collection of specimens in Natural History, and having forwarded to the Smithsonian Institution, large quantities, embracing multitudes of new species and the types of several new genera, that Institution offered to deposit a suite of his specimens in such seat of learning as Lieut. Trowbridge might select. He selected the University of his native State; and, in accordance with the promise made, the Smithsonian Institution supplied, not only a selection of specimens from Lieut. Trowbridge's collections, but added hundreds of specimens from its miscellaneous stores. During the winter of 1858–9 Mr. R. Kennicott was employed by the University to select the specimens which were to constitute the collection; and in April, 1859, the collection arrived and was deposited in the new cases of the zoological gallery. A catalogue of this collection was published by the University in 1861.

The following is a statement of the contents of the "Trowbridge Collection."

| | | | | |
|---|---|---|---|---|
| North American Mammals, - - - | 28 | genera, | 56 | species. |
| " " Birds, - - - - | 152 | " | 240 | " |
| " " Reptiles (including Batrachians,) | 73 | " | 165 | " |
| " " Fishes, - - - - | 74 | " | 95 | " |
| Exotic Mammals, - - - - - - - | 8 | " | 8 | " |
| " Birds, - - - - - - - - | 7 | " | 10 | " |
| " Reptiles, - - - - - - - | 8 | " | 14 | " |
| Coleoptera, (Beetles,) - - - - - | 83 | " | 111 | " |
| Orthoptera, (Grasshoppers, &c.,) - | 19 | " | 26 | " |
| Neuroptera, (Dragon Flies, &c.,) - | 16 | " | 36 | " |
| Lepidoptera, (Butterflies, &c.,) - - | 26 | " | 47 | " |
| Hemiptera, (Squash Bugs, &c.,) - | 51 | " | 65 | " |
| Homoptera, (Tree Hoppers, &c.,) - | 18 | " | 36 | " |
| Diptera, (Flies, &c.,) - - - - - - | 63 | " | 99 | " |
| Myriapoda, (Centipede,) - - - - - | 1 | " | 1 | " |
| Arachnida, (Spiders, &c.,) - - - - | 3 | " | 4 | " |
| Crustacea, - - - - - - - - - - | 46 | " | 64 | " |
| Mollusca, - - - - - - - - - - - | 19 | " | 39 | " |
| Totals, - - - - - - - - | 695 | " | 1204 | " |

The whole number of entries in the collection is 1356, showing that 152 entries are of duplicate specimens from different localities.

The specimens in this collection possess unusual value, in consequence, 1st, of the remoteness and difficulty of access of the regions in which they were, for the most part, collected; and 2d, from the fact that the various orders were labeled by gentlemen who are experts in their several fields of study. Most of the mammals, birds and reptiles, passed directly under the scrutiny of Prof. S. F. Baird, in the preparation of his unequalled Reports on the Mammals, Birds and Reptiles of North America.

Geological Survey of 1859–60. By the terms of the law re-organizing the scientific survey of the State, one suite of all the specimens collected was to be deposited in the University.* The field work of the survey has been prosecuted but two seasons; and during this time the collection of specimens has had regard rather to the illustration of a rapid *reconnoissance*, than to the accumulation of a store of materials for distribution. Nevertheless, the products of the survey embrace specimens of rocks and fossils from all the formations occurring south of the Sault St. Mary, among which are very many species new to science—one hundred and fifty of which have recently been described by the writer while descriptions of others are soon to be published. They embrace also a pretty complete series of the fishes of the Maumee river, and the streams of Monroe County, with many of the reptiles and batrachians. The other zoological specimens of the survey—of which the University is entitled to at least one suite—were removed to the Agricultural College, and have not yet been recovered. The products of the survey embrace also 274 species of plants not enumerated in Dr. Wright's botanical report of 1838.

It is greatly to be regretted for the interests of the University, no less than for those of science and the State itself that this survey could not have been continuously prosecuted; especially since appropriations made by two successive legislatures still remain unexpended.

W. S. Woodruff, an alumnus, presented in 1860, a box of fossils collected by himself from the Eocene beds, at Claiborne, Ala.

William Mendenhall, of Richmond, Ind., a graduate of 1863, presented, in 1860, a small lot of lower Silurian fossils from the

*Laws of Michigan 1859, p. 565.

vicinity of Richmond; and, in 1863, a few specimens from the iron region of southern Ohio.

HOYT POST, a graduate, presented in 1860, 482 specimens of insects collected in the vicinity of Detroit.

JAMES H. GOODSELL, a graduate, presented 331 specimens of insects from Oakland County, collected in 1860.

A. H. WILKINSON collected also in 1860 112 specimens of insects for the University.

ALBERT D. WHITE, late of Delaware County, Ohio, deposited, in 1858–9–60, sundry osteological preparations, reptiles, and other specimens; and in 1861, a collection of fossils from the Corniferous limestone of Columbus, Ohio. After Mr. White's lamented death in the army, in March, 1863, his father, in fulfillment of what he conceived would have been his son's desire, presented these specimens to the Museum of the University. Mr. White was a modest and disinterested votary of science, and though laboring under heavy embarrassments, had already laid the foundations of an honorable distinction, had not the casualties of war borne him, too soon for us and for science, to another sphere of existence.

JOSEPH BROWN, ESQ., of Ann Arbor, contributed in 1860, a large block from his "marble quarries" at Lamont, Illinois, containing a "pot hole" excavated probably by running water, at a period when the valley in which the quarry is situated was the bed of a river giving outlet to Lake Michigan. In 1862, Mr. Brown presented samples of the ores and rocks from the lead region of Wisconsin.

LEWIS MCLOUTH, a graduate of the University presented, in 1862, a box of geological specimens from the copper mines of Lake Superior. Many of these specimens were contributed by Alfred Meads, Esq., of Ontonagon.

LIEUT. C. H. DENISON, an alumnus, presented in 1862 some rock specimens from near Mount Vernon, Va.; also a serpent from the same vicinity.

PROF. A. DUBOIS presented in 1862 two large, fine specimens of the King Crab from Gravesend, L. I.; also a *Platycarcinus Sayi* and a few fishes; and, in 1863, two specimens of *Pyrula canaliculata*, and a large Star Fish.

THE ACADEMIC CLASS OF 1862 deposited on the University

grounds a huge rounded boulder of jaspery conglomerate weighing about six tons. This was found projecting slightly above the surface in a city lot near the rail road depot, and was raised by the enterprise of the class and removed on a sledge in February. The nearest locality at which such a conglomerate is known to be in place, is the north shore of Lake Huron, in the neighborhood of the Bruce copper mines. The class have left their inscription on the flat side of the boulder; and it thus becomes a monument to their memory which will as far outlast the pyramids as it already exceeds them in antiquity.

Charles A. White, of Burlington, Iowa, sold the University in 1863, for $500, his fine geological and zoological collection, consisting of 1223 entries and about 6000 specimens. The geological specimens embrace 1018 entries and the zoological 205. This collection is remarkable for two things: 1st, the large number of beautifully preserved crinoids which it contains, and 2d the number of its original or type specimens The fossils in this collection probably double the number previously in the possession of the University. The "White Collection" is described in greater detail in a special report appended to the present one.

Mr. White has more recently donated a small collection of fossils from Burlington, Iowa; and still later a large box of geological specimens from the carboniferous limestone at Burlington, and the Hamilton group of other parts of Iowa. This box contained also two serpents, a centipede and a laceratian from Texas.

Rev. D. C. Jacokes presented this summer a collection of skulls; consisting of five skulls of human races, and six skulls of carnivorous quadrupeds.

Joseph W. Wood, a graduate of 1861 and of 1862, presented, in July last, a small collection of extremely interesting fossils from the lowest horizon of life in Wisconsin. A notice of these fossils with descriptions of the new species is now in process of publication in Silliman's Journal.* The new species are *Orthis Barabuensis*, *Straparollus (Ophileta) primordialis*, *Pleurotomaria advena*, *Ptychaspis Barabuensis*. Accompanying these fossils were about 15 species of Unionidæ from Wisconsin.

*See Sill. Jour. for March, 1864.

Smaller donations to the museum have been made by a large number of persons, of whom I am able to enumerate the following:

John Schneeberger. Section of an oak tree with a pair of deer's horns overgrown, Washtenaw County, 1856.

John Vance. *Chrysemys marginata Ag.*, Ann Arbor, 1856.

Prof. A. B. Palmer. Eggs of sand hill crane. Illinois, 1856.

Hon. H. P. Van Cleve, Ann Arbor. Skull and horns of Elk, Minnesota, 1857.

J. McCrary. Skin of *Lynx canadensis*, Portland, Ionia County, 1857.

Dr. C. B. Porter. Skin of *Taxidea Americana.* 1857.

J. E. Challis. Fœted limestone inclosed in the "Black shale," Kettle Pt., (Cape Ipperwash) C. W. 1857.

J. A. Brown. Bones from an Indian tomb, Canada West. 1857.

E. E. Baldwin. Bones from an Indian tomb, Ann Arbor, 1858. Also the skin of an opossum captured at Ann Arbor—a very rare species so far north.

S. G. Morse. Ores of iron from Marquette, Mich. 1857.

Mrs. F. J. B. Crane. Two masses of shell conglomerate from Paradise I., Florida. 1857.

W. C. Voorhies. Fossil wood from drift, Ann Arbor, 1857.

T. M. Easton. Slab of whalebone 10 feet in length, from Shanta Bay, Kamtschatka. 1858.

J. Q. A. Fritchey. Truncheons of wood from Ann Arbor and St. Louis, Mo. 1857.

Hiram Parker. *Lepidodendron* from near Mason, Ingham county. 1858.

Hon. L. H. Parsons. Specimens of Coal, Corrunna, Mich. 1858.

W. S. Abbey. "Saw" of Saw fish from Indian river, Florida. Presented through Hon. W. S. Maynard, 1858.

T. Howard. Weathered mass of hornblendic granite.

H. E. Kingsbury. Distal extremity of femur of *Mastodon giganteus.* Nebraska, 1858.

M. VANDERCOOK. Double-headed Snake. (*Heterondon platyrhinos*) from Livingston Co.

O. B. WHEELER. Mass of crystalized copper from Lake Superior. 1858.

J. ELISHA WINDER. Fragment of auriferous quartz from California.

PROF. C. L. FORD. Silicified wood from Bladensburgh, Md., and fossil shell from Mt. Vernon, 1858.

D. DEPUY. Mass of fossiliferous limestone (Corniferous), Pittsfield, Washtenaw Co. 1859.

F. A. CHASE. Fossils from the Hamilton group, N. Y.

R. HOOPER. Boulder of native copper from a well. Ann Arbor, 1859.

MRS. A. M. REDFIELD. A mass of *Rhynchonella plena* and a collection of *Unionidæ* from L'Original, C. W.

CAPT. G. TRAVERSE. Iron Ore, Marquette, 1860.

MRS. H. W. LEAVENWORTH. "Horned Frog." Texas, 1860. This reptile was sent alive by mail, in a tin box.

S. S. WALKER. Septaria, ("Turtle Stones") from Fredonia, N. Y. 1860.

PROF. H. S. FRIEZE. Seventeen years locusts. Staten Island. 1860.

MRS. DR. TICKNOR. Cannel coal and sandstone, 50 miles from Cleveland, Ohio, 1860.

M. N. HALSEY. Verd antique (ophiolite) 18 miles from Montpelier, Vt., 1860.

LEWIS W. WOOD, M. D. Fossil shell, (*Ambonychia neglecta*) from Chicago, Ill.

A. J. CHAPMAN. Drift fossils. Ann Arbor, 1860. Geological specimens from Kentucky. 1861.

HIGBY & STEARNS. Yellow Catfish and young Rock Sturgeon. Detroit river, 1860.

D. SPERRY. Carapace of snapping turtle (*Chelydra serpentina*). Ann Arbor, 1860.

Col. Castle Sutherland. Indian stone hatchet. Genesee Co. 1860.

C. E. Hovey. Fine specimen of Grand Rapids gypsum. 1860.

S. M. Billings. Wallastonite, Minnesota mine L. S. 1861.

David E. Ainsworth. Siamese account book. Bankok, Siam. 1861.

J. B. Blenkiron. Centipede, said to be from England. 1862.

R. J. Sloan. Jaw of fossil fish, (*Macropetalichthys Sullivanti*) from corniferous limestone of Maitland river, C. W. 1862.

G. B. Tichenor. A new species of gar pike *Lepidosteus oculatus*, Winchell,) from Duck Lake, Calhoun County, 1863.

Mrs. A. M. Edwards. Fragment of chimney swallow's nest built in horizontal portion of stove pipe. Ann Arbor. 1863.

Rev. George Taylor. Silicified wood, and mass of sandrock, containing casts of *Turritella Mortoni*, Fredericksburg, Va. 1863.

Dr. C. Rominger. Fossils from the Marshall sandstone, Marshall. 1863.

W. H. Boothroyd. Samples of salt manufactured at Saginaw by the various methods. 1863.

Fayette Hurd. Truncheon of Persimmon wood. Monticello, La. 1863.

David G. Colwell. Tooth of Mastodon. Tyrone, Livingston County. 1863.

A. Drury. Efflorescence of sulphate of iron from a peat bog. Pittsfield, Washtenaw County. 1863.

Mrs. George B. Russell. A weathered boulder of syenite intersected by a projecting seam of quartzite. 1863.

Dr. Rixsecker. Ores of iron and zinc. Bethlehem, Pa. 1863.

C. F. Conrad. Travertin from T. 27 N. 9 E., and marl from 4½ miles north west of Alpena, 1863.

A. D. Peck. Travertin and Polishing Powder (good) from Sec. 12, Hamburgh, Livingston County.

Col. O. L. Mann. Sections of Palmetto wood from Morris Island, S. C. 1863.

According to the foregoing enumeration, the materials of the collections of Geology, Zoology and Botany* have been derived from 93 distinct sources, the names of which are preserved in the records of the museum or upon the labels of the specimens. It is quite likely that the authors of some donations have been overlooked, in consequence of the former defective method of keeping the records.

The books employed in the museum are four.

1. A *Journal* in which are entered the date of every acquisition, its nature, and the source whence obtained.

2. A *Geological Register* in which the names of specimens are written opposite the serial numbers extending from one upwards; and opposite these, the locality, formation, source of acquisition, date of acquisition, collector and place in the museum of each specimen. The specimens thus registered are labeled and disposed in the cases in two different ways. All small specimens are firmly and permanently mounted on cards handsomely printed, bearing the register-number, name, locality, formation and name of collector of the specimen. Larger specimens are allowed to rest on the shelves with the labels lying in contact with the specimens. On each specimen is securely fastened a small circular piece of white paper bearing the register number, which is also on the label. By this, the label and specimen are kept associated, and in case of loss of label, it is easily restored by reference to the proper number in the Register. Specimens obtained from the Scientific Survey of 1859-60, are so indicated by an entry on the card on which they are mounted; or, in other cases by a yellow oval piece of paper firmly glued to the specimen, and bearing the locality number of the survey. In the general collection, the specimens are arranged chronologically under their several geological groups—the newer groups being always above and toward the left. Within the limits of a group, fossils are arranged zoologically—the higher species being above and to the left.

3. A *Zoological Register* giving the serial number, original number, name, sex, locality, date of collection, nature of specimen, measurements, source whence obtained, collector, number of specimens and remarks.

---

*A few donations of an ethnological character are embraced since we have no distinct department for specimens of this kind.

4. A Register of the Trowbridge Collection. This, since the unconditional donation of the collection to the University, has become obsolete.

It is not possible, in the present state of our collections, to arrive at an accurate idea of their extent. The Botanical Cabinet is composed nearly as follows:

| | Species. | Entries. | Specimens. |
|---|---|---|---|
| Plants of the survey of 1837–8, | 900 | 1200 | 8000 |
| " " " 1859–60, | 274 | 300 | 1000 |
| Truncheons of Wood, | 30 | 30 | 35 |
| Totals, | 1204 | 1530 | 9035 |

The Zoological Cabinet is made up nearly according to the following statement:

Table showing the number of Entries and the number of Specimens in each of the Classes of the Zoological Cabinet of the University of Michigan, September 30th, 1863.

| CLASSES. | Survey 1837–8 | | Monds Collec'n. | | Trowbr. Collec'n. | | Smith'n Inst'tn. | | Survey 1859-60 | | White Collec. | | Other Sources. | | TOTALS. | |
|---|---|---|---|---|---|---|---|---|---|---|---|---|---|---|---|---|
| | Entries | No. Spec. | Entries. | No. Spec | Entries. | No. Spec. | Entries. | No. Spec. | Entries. | No. Spec. | Entries | No. Spec | Entries. | No. Spec. | Entries. | No. Spec. |
| Mammals, | 25 | 30 | .... | .... | 66 | 66 | 28 | 28 | .... | .... | ... | ... | 12 | 20 | 131 | 144 |
| Birds, | 241 | 748 | .... | .... | 376 | 380 | 147 | 147 | .... | .... | ... | ... | 16 | 16 | 780 | 1291 |
| Reptiles, | 20 | 40 | .... | .... | 232 | 245 | .... | .... | 7 | 13 | ... | ... | 31 | 38 | 290 | 336 |
| Batrachians, | 8 | 20 | .... | .... | 51 | 64 | .... | .... | .... | .... | ... | ... | 22 | 128 | 81 | 212 |
| Fishes, | ... | .... | .... | .... | 96 | 125 | .... | .... | 52 | 217 | ... | ... | 24 | 28 | 172 | 370 |
| Insecteans, | ... | .... | .... | .... | 429 | 550 | .... | .... | .... | .... | ... | ... | 188 | 850 | 617 | 1400 |
| Crustaceans, | ... | .... | 4 | 4 | 67 | 103 | .... | .... | 6 | 40 | ... | ... | 20 | 40 | 97 | 187 |
| Cephalopods, | ... | .... | 8 | 13 | .... | .... | 1 | 2 | .... | .... | ... | ... | .... | .... | 9 | 15 |
| Gasteropods, Marine, | ... | .... | 1205 | 3612 | 17 | 28 | 167 | 205 | .... | .... | ... | ... | 15 | 30 | 1404 | 3785 |
| " Fluviatile, | ... | .... | 68 | 322 | .... | .... | .... | .... | 7 | 28 | 23 | 28 | 7 | 48 | 105 | 486 |
| " Pulmoniferous, | 18 | 246 | 235 | 1219 | .... | .... | 8 | 22 | 19 | 655 | 50 | 200 | 72 | 304 | 402 | 2646 |
| Lamellibranchs, Marine, | ... | .... | 293 | 663 | .... | .... | 32 | 39 | .... | .... | 128 | 315 | 20 | 50 | 473 | 1067 |
| " Fluviatile, | 20 | 44 | 17 | 94 | 16 | 20 | 120 | 120 | 20 | 44 | ... | ... | 103 | 194 | 296 | 516 |
| Brachiopods, | ... | .... | 2 | 6 | .... | .... | .... | .... | .... | .... | ... | ... | 1 | 6 | 3 | 12 |
| Echinoderms, | ... | .... | .... | .... | .... | .... | .... | .... | .... | .... | 1 | 1 | 10 | 20 | 11 | 21 |
| Polypi, | ... | .... | 8 | 8 | .... | .... | .... | .... | .... | .... | 2 | 2 | 4 | 10 | 14 | 20 |
| Totals, | 332 | 1128 | 1840 | 5941 | 1350 | 1581 | 503 | 563 | 111 | 997 | 204 | 606 | 545 | 1782 | 4885 | 12598 |

The Geological Cabinet is not as yet in such a condition as to admit of an analysis of its contents or even more than a rough estimate of its aggregates. The following figures propably fall below the truth:

| | Entries. | No. of Spec. |
|---|---|---|
| From Survey of 1837–8, - - - - | 3000 | 9000 |
| " " 1859–60, - - - - | 3000 | 11000 |
| White Collection, - - - - | 1018 | 5394 |
| Other Sources, - - - - - - | 250 | 650 |
| Totals, - - - - - | 7268 | 26044 |

The following is consequently a rough estimate of the entire collections in Geology, Zoology and Botany:

| | Entries. | No. Spec. |
|---|---|---|
| In Botany, - - - - - - | 1530 | 9035 |
| In Zoology, - - - - - - | 4885 | 12598 |
| In Geology, - - - - - - | 7268 | 26044 |
| Grand Aggregates, - - - | 13683 | 47677 |

The requisites of a museum for purposes of education are far from answered when a large amount of rough material has been brought together. Such material requires to be arranged, labeled and properly exhibited in cases suited to the different classes of objects. Before the arrangement can be entered upon, much labor is often requisite in preserving, dressing or working out the specimens. This is especially the case in Geology. The rough specimens as taken from the field are generally quite unfit for the cabinet. In regard to fossils it may be stated that the working geologist often spends hours in working out from the surrounding rock, a single shell, so as to exhibit advantageously all parts of its structure. With crinoids, two or three days are often devoted to the cleaning up of a single good specimen.

Next, after a specimen has been preserved, and, if a fossil, thoroughly worked out of its matrix and cleaned, is the determination of its name, if described, and lastly, its mounting and appropriate exhibition. Now, it is well known that the University, previous to the purchase of the White Collection was in possession of very few geological specimens not the product of the public surveys of the state. These materials have existed in a crude condition. In the next place, the peculiar position of our peninsula, has caused it to enjoy a geological history of its own; so that most of the species found fossil within our boundaries prove to be entirely new to science. Thus, after the most faithful search for the identification of a species, we find, at last, that it is entirely unnamed; and are obliged to draw up and publish a description of it for ourselves. Happy is the investigator, if, in this labor, he finds in our meagre library, such works as are indispensable to a palæontological investigation.

In consequence of the great amount of labor and time involved in these researches, the final working up and mounting of the geological specimens proceeds but slowly. We have, however, attained a vantage ground, by the study which has already been devoted to these specimens, which will enable us to advance with increased rapidity.

In the Zoological Cabinet the work is more advanced; and, excepting a portion of the marine shells and a portion of our domestic fishes, there are few specimens not determined. Unfortunately the determination of the shells is utterly impossible with such facilities as our library affords. The same may be said of most of our small fishes. More than this, it is impossible that one individual should make so much of a speciality of half a dozen classes of animals as to acquire the ability and authority to label specimens belonging to all of them. It is scarcely possible for one devoting his whole time to *Insects alone*, to make determinations of specimens in the various orders of that single class. What shall we say when to insects are added shells, fishes, reptiles, birds, mammals, plants and the whole vast field of geology?

Changes and additions are required in the cases for the suitable accommodation of the various classes of specimens.

1. Our fine collection of shells has not a single case suited to its wants. I have heretofore made recommendations on this subject which have not been adopted; but changes now in contemplation, will, perhaps open the way to other suitable provision.

2. No suitable means exist for the protection and display of the cases of insects. A very small sum would remedy this deficiency.

3. A good case is needed for a collection illustrative of economical geology.

In addition to these wants it may be doubted whether it is the wisest policy to postpone any longer the mounting of some of the best skins of birds and quadrupeds now lying in the zoological cases.

I would also recommend, as heretofore, the printing of such catalogues as are ready, and would remind the Board of the desirableness of some general and uniform plan for the style of all our printed museum catalogues.

The following is a list of the catalogues in my department which are quite ready for printing, or as nearly so as is practicable until they are actually wanted by the printer:

| | | |
|---|---|---|
| 1. Plants, | 7. Insecteans, | 12. Potsdam Group. |
| 2. Mammals, | 8. Crustaceans, | 13. Huron " |
| 3. Birds, | 9. Unionidæ, | 14. Chemung " |
| 4. Reptiles, | 10. Helicidæ, | 15. White Collection. |
| 5. Batrachians, | 11. Other Mollusca as far as determinable. | 16. Lithological Cabinet. |
| 6. Fishes, | | |

I desire to add a few words of suggestion as to the future disposition of our specimens. It is stated that the Mineralogical Cabinet requires more space. I see no objection to devoting the entire gallery to its accommodation. It is certain however that the adjoining room cannot accommodate the whole of the geological collection already in our hands. The space now occupied by the general library would do so, and would form a suitable depository and place of exhibition. In the mean time the room now occupied by geology can be devoted to a department of ethnology and archæology—a department in which we already possess many interesting illustrations, and which might well be made a speciality by the only great University that has been reared on the ancient hunting grounds of the Indian tribes of the Northwest. Lastly, floor-cases could then be constructed for the shells, and they could be handsomely arranged in a solid body on the floor now occupied by the library.

Our museum is rich in certain departments; as for instance, Mammals, Birds, Reptiles and Molluscs—as also Crinoidea, and fossils from the Chemung group of rocks. It now has a respectable representation of the Faunas of the Palæozoic formations of North America; and, in fine the whole geology of the lower Peninsula of our state is quite amply illustrated. What we want further is a suite of specimens illustrating the general features of European geology. This would consist mainly of a collection of European fossils, which have, of necessity, always been made the standards with which American species have been compared. Especially must we look to European sources for means to illustrate most of the formations of the Mesozoic system.

To add eclat, and inspire a livelier interest in geology, some specimens of greater magnitude and more striking characters should be obtained. A cast of the *Megatherium* or *Deinotherium* would strike the imagination and inspire an interest which specimens of less magnitude would never awaken.

ALEXANDER WINCHELL,
Prof. Geol. Zool. and Bot.

UNIVERSITY OF MICHIGAN,
Ann Arbor, 30th Sept. 1863.

# REPORT ON THE WHITE COLLECTION.

To the Honorable, the Board of Regents of the University of Michigan:

The undersigned, having completed the arrangement, mounting, labeling and registering of the collection of geological and zoological specimens purchased of Charles A. White, is prepared to furnish the following more detailed statement of the nature and extent of the collection, and its value as an acquisition to the Museum of the University.

The entire collection consists of 1223 entries in the Registers, and about 6000 separate specimens (exclusive of the countless numbers of some small species) every one of which has been examined and assigned its proper place in the arrangement. The collection embraces 1018 geological entries and 205 zoological. Of the geological entries 955 are American species and 63 European. Of the American species all but 16 entries are well preserved fossils from the rocks of the northern states—if we except 17 species from the Tertiary of Alabama.

The following is a more detailed exhibit of the constitution of this valuable collection:

### ANALYSIS OF THE WHITE COLLECTION.

#### *A. Geological Specimens.*

I. American Specimens.

(a) Fossils.

| | |
|---|---|
| Trenton Group, - - | 13 |
| Hudson River Group, - - | 38 |
| Niagara Group, - - - | 65 |
| Clinton " - - - - | 6 |
| Lower Helderberg Group, - - | 41 |

| | | | |
|---|---|---|---|
| Oriskany Group, | | 8 | |
| Upper Helderberg Group, | | 18 | |
| Hamilton Group, | | 120 | |
| Chemung Group, | | 152 | |
| Burlington Limestone, | | 242 | |
| Crinoids, | 212 | | |
| Other Families, | 26 | | |
| Keokuk Limestone, | | 57 | |
| Crinoids, | 28 | | |
| Other Families, | 29 | | |
| Warsaw Limestone, | | 61 | |
| Crinoids, | 5 | | |
| Other Families, | 56 | | |
| Kaskaskia Limestone, | | 37 | |
| Crinoids, | 9 | | |
| Other Families, | 28 | | |
| Coal Measures, | | 35 | |
| Tertiary, | | 17 | |
| Post Tertiary, | | 3 | |
| Remaining to be investigated, | | 30 | |
| | | — | 939 |

(b) Rocks.

| | | |
|---|---|---|
| Azoic, | 3 | |
| Trenton, | 2 | |
| Clinton, | 1 | |
| Hamilton, | 1 | |
| Chemung, | 1 | |
| Burlington Limestone, | 4 | |
| Coal Measures, | 1 | |
| Permian, | 1 | |
| Post Tertiary, | 2 | 16 |
| | — | — |
| Total American Specimens, (entries) | | 955 |

II. European Specimens.

(a) Fossils.

| | | |
|---|---|---|
| Silurian, | 9 | |
| Devonian, | 3 | |
| Lias, | 8 | |
| Cretaceous, | 5 | |
| Oolite, | 1 | |
| Tertiary, (Paris Basin) | 36 | |
| Post Tertiary, | 1 | |
| | — | |
| Total European (entries) | | 63 |
| Total entries in the geological section, | | 1018 |

*B. Zoological Specimens.*

| | |
|---|---|
| Radiates, - - - - - | 3 |
| Unionidæ - - - - | 123 |
| Cycladidæ, - - - - - - | 5 |
| Paludinidæ - - - - | 23 |
| Lymneidæ, - - - - - | 15 |
| Helicidæ, - - - - | 35 |

Total zoological entries, - - - - - 205

Grand Total in the White Collection, - - 1223 entries.

The collection possesses a few peculiarities which render it unique, and one which it will be impossible to duplicate. One of its striking features is its richness in Crinoidea. Of this order it embraces not less than 280 entries, all or nearly all of which belong to the tribe of Crinids or Crinoidea possessing stems and arms. The value of this department of the collection will be better appreciated when it is remembered that Dr. Bronn in his "*Index Palæontologicus*," published in 1849, enumerates but 314 well determined species of Crinoids then known from all parts of the world, while only 250 of these were armed and pedunculate—being 30 less than is contained in the White Collection alone. The richness of the collection in this order of extinct animals is set in a still stronger light by the fact 264 entries are Crinoids from the Carboniferous Limestone, a formation from which Bronn enumerates but 64 species from all parts of the world. The writer ventures the assertion that there is not another such a collection of Carboniferous Crinoidea in existence. What is still further remarkable is the fact that 212 species of the Crinoids of the collection come from the "Burlington Limestone," or lowest member of the great carboniferous limestone formation. These were collected at the typical, and at the same time, the most favorable outcrop of these strata; and consist of specimens which the demuding agencies of thousands of years had been working from the rock, and which it will require thousands of years more to reproduce in equal abundance, perfection and beauty.

In fossils of the Chemung Group the collection is also surprisingly rich. The number of entries is 152, while the total number of species known is not much over 360. Of the known species not in the White Collection, nearly one half are New York types, thus far imperfectly described, and for the most part existing only in the State

Collection at Albany; while 91 other species come from Michigan and are also in the possession of the University; so that we may say there is scarcely a well known Chemung fossil which is not represented in our museum.

Another most important feature of the collection is the number of original or type specimens which it embraces. Of the whole number of American species, 232, or more than one fourth, are the identical specimens from which the original descriptions were drawn up; and hence must always possess an authenticity entirely unique. Figures of not a few of these specimens have found their way into the elementary books of the science. To this number may yet be added many of the uninvestigated specimens before alluded to.

It may add still further to an exposition of the value of this acquisition to compare it with the "Trowbridge Collection" in Zoology, which has been so highly estimated by the authorities at the Smithsonian Institution. This collection consists of 1356 entries, and perhaps 3000 specimens. The Trowbridge Collection is justly valued in consequence of its possession of a large number of "type specimens." It is true that most of the birds and mammals bear Smithsonian numbers which are quoted in Prof. Baird's works on the Ornithology and Mammalogy of North America; and the specimens undoubtedly passed under his scrutiny during his investigations. But it is folly to suppose the most perfect and authentic types were not retained in the Smithsonian Collection. Those in our possession were truly labeled by the highest authority— by original authority or under its cognizance. They may have served as types in part but we can scarcely style them *the* types of the several species. The types of the White Collection correspond to those perfect specimens reserved for the Smithsonian Institution. They are the sole types of the several species; and no other collection in existence can ever claim the possession of specimens carrying with them equal authority. It must not be regarded as invidious, if we still more emphatically allude to the excellence of the White Collection in the beauty and perfection of its specimens. It is not a selection from among the refuse or even the duplicates of a larger collection, with the list of numbers made out by an application of that maxim of true science but of dernier resort, that the most defective examples are desirable till better are obtained. The specimens here are the best in existence.

The University, therefore, may be very justly congratulated, as it is, by the proprietors of other collections, in having so promptly availed itself of one of the most desirable cabinets in the United States.

There would be a propriety in the attempt to preserve to some extent the separate identity of this collection. Such distinct identity would most effectively encourage the former owner to execute his generous purpose of making this cabinet the permanent receptacle of his future discoveries. The execution of such purpose has been already commenced in the transmission of some new fossils of extreme interest, among which may be mentioned an oyster dating back a whole geological epoch anterior to any oysters hitherto known.* The most effectual method of preserving the distinct identity of the cabinet would of course be to mount it in a separate case. Next to this, a separate catalogue, as has been done with the Trowbridge Collection, would insure its independent existence in the history of the museum. All things considered this is probably the most expedient course. Lastly, not a little is already effected toward the end referred to, by the credit given to the collection on the cards which mount the specimens, and in the geological and zoological Registers of the museum. This credit should, in any event, be scrupulously preserved in any geological catalogue which may hereafter be published.

Respectfully submitted,

A. WINCHELL,

Prof. Geol. Zool. and Bot.

UNIVERSITY OF MICHIGAN, 26th Sept. 1863.

---

*Since this was written Mr. White has donated a large box of geological specimens from Iowa.

www.ingramcontent.com/pod-product-compliance
Lightning Source LLC
LaVergne TN
LVHW011141110826
845150LV00008B/2457

* 9 7 8 1 4 1 8 1 9 2 9 0 7 *